Mosses & Liverworts

naturally scottish

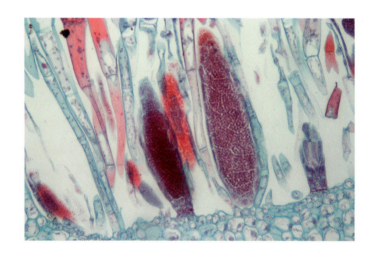

© Scottish Natural Heritage 2005

ISBN 1 85397 446 3 paperback

A CIP record is held at the British Library

NP1K0605

Acknowledgements:

Advice and comments from David Long

Author: Gordon P. Rothero

Series Editor: Lynne Farrell (SNH)

Design and production: SNH Design and Publications

Photographs:

David Holyoak 25; **Lorne Gill/SNH** front cover, opposite 1, 9, 15, 19, 24 bottom right, 27, 31 top, 31 bottom, 34 bottom right, 35; **John MacPherson/SNH** 28 bottom; **Steve Moore/SNH** 22, 28 top; **Gordon P. Rothero** back cover, contents, 1, 2, 5 top left, 5 top right, 5 bottom left, 5 bottom right, 11 top left, 11 top right, 11 bottom left, 11 bottom right, 13 top, 13 bottom, 16 top left, 16 top right, 16 bottom left, 16 bottom right, 17, 18, 21 top left, 21 top right, 21 bottom left, 21 bottom right, 24 top left, 24 top right, 24 bottom left, 26 top, 26 bottom, 33, 34 top left, 34 top right, 34 bottom left, 36; **Bruce Westling** frontispiece.

Scottish Natural Heritage
Design and Publications
Battleby
Redgorton
Perth PH1 3EW
Tel: 01738 458530
Fax: 01738 458613
E-mail: pubs@snh.gov.uk
www.snh.org.uk

Cover photograph:
Typical wet patch with Common haircap *Polytrichum commune* and Cow bog-moss *Sphagnum denticulatum* and the liverwort Water earwort *Scapania undulata*

Frontispiece:
Section through male antheridia (male organ of a moss)

Back cover photograph:
Broad-nerved hump-moss *Meesia uliginosa* growing with Dwarf willow

Mosses and Liverworts

naturally scottish

by

Gordon P. Rothero

Foreword

When Scotland emerged from its mantle of ice, it had been scoured bare, leaving rounded hills, jagged mountains, deep valleys, undulating lowlands, lakes and wetlands. All were ripe for colonisation by plants, and the group best equipped to move in quickly were the most ancient of the land plants, the bryophytes - mosses, liverworts and hornworts - pioneers of rock, soil and water, which are perfectly adapted for rapid dispersal by their vast numbers of tiny wind-borne spores. The diversity of Scotland's geology and the cool, wet climate favoured rapid colonisation by a wide range of bryophytes, and as trees became increasingly dominant, so the bryophytes diversified further. The long history of man's destructive activities has brought about loss of both mossy habitats and species, but has also produced new habitats and new incomers from overseas, right up to the present day.

As a result, bryophytes are amongst the most dominant and beautiful components of many of Scotland's habitats, perhaps most spectacular and luxuriant in the oceanic ravines and oakwoods of the western seaboard (our own temperate rainforests) and in the unique liverwort heaths of the north-western mountains. They are also conspicuous components of many other habitats from raised mires to montane springs and flushes, coastal sand dunes and slacks with their esoteric rarities, to rich boulder screes and mossy cliff ledges throughout the uplands. On a global scale our bryophyte flora is truly outstanding in its wealth of species and luxuriance, and every corner of the country, from the wind-blasted cliffs of St Kilda to the summit of Ben Nevis, and to the warm sea-banks of the Black Isle and Berwickshire, has its own specialities.

Yet bryophytes for all their aesthetic appeal and ecological importance have often been ignored and neglected, new discoveries are still regularly made by careful field work and taxonomic research, and long-overdue measures to protect this unique heritage are at last being implemented. This new publication, by one of the country's foremost bryologists, will surely help to increase awareness of the beauty and importance of this heritage, attract new converts to their study, promote exploration of parts of the country still unsurveyed, and encourage recognition of our responsibilities to protect our rich diversity of species and habitats.

Bryophytes, like their frequent consorts the lichens, are highly sensitive to air and water pollution, and many 'indicator species' found in Scotland demonstrate the high environmental cleanliness of our wild places. However, this sensitivity makes them very vulnerable to man-made changes such as eutrophication and climate change. We must be ever-vigilant in monitoring the tundra bryophytes of the Cairngorm snow-beds, carry out essential research such as study of the spectacular disjunct species found only in Scotland and the Himalayas or South America, and be diligent in opening the eyes of schoolchildren to the wonders of the secret world of bryophytes. Gordon Rothero and Scottish Natural Heritage have, in this book, made a splendid contribution to this process.

Dr David G. Long
Bryologist, Royal Botanic Garden, Edinburgh

Rock fingerwort *Lepidozia cupressina* - a fine liverwort of rocky western oakwoods

Contents

Common haircap *Polytrichum commune* - one of our largest mosses

Mosses and liverworts in Scotland

Mosses and liverworts are everywhere in Scotland, adding colour and interest to both the most mundane and the most extreme of habitats. In the woodlands, in the mires and on the mountain tops where they are most abundant, the collective texture and pattern of the many species is stunningly beautiful. Mosses and liverworts (collectively called 'bryophytes') and the plant communities of which they are major constituents, are more frequent and more diverse in Scotland than in any other part of the UK and in most other parts of Europe. Some of these bryophyte communities, particularly those of the western 'rain forests' and the 'oceanic heaths', are globally very rare and have species which are limited to a few, often geographically distant, localities.

In Scotland there are just under 1,000 species of moss and liverwort, some 87% of the UK total. In international terms, Scotland has more than 60% of the European bryophyte flora, including some endemics such as Scottish beard-moss *Bryoerythrophyllum caledonicum* and Scottish thread-moss *Pohlia scotica*.

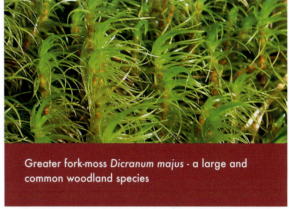

Greater fork-moss *Dicranum majus* - a large and common woodland species

With estimates of the total number of global species ranging from 16,000 to 24,000, it is possible that Scotland may have as much as 5% of the world's bryophytes. These are astonishing figures, considering the country's small size and the relative poverty of our flowering plants using the same comparisons.

If you wander into any of the remnants of oak and birch woodland on the west side of Scotland, particularly in Argyll, in the area around Loch Sunart and north to Sleat on the Isle of Skye, you will see that the trees and rocks are covered with a deep carpet of mosses and liverworts, often extending up into the canopy. Closer examination will reveal that it is not all the same plant but a whole range of species, each with its own niche.

1

In the extensive areas of mire for which Scotland is so important, bryophytes can be even more abundant, certainly in terms of sheer biomass if not in species diversity. The mossy lawns and hummocks in the mires are composed of bog-moss, species of *Sphagnum,* and the health of these mires is largely dependent on the well-being of the *Sphagnum.* The dead but only partially decayed remains of these bog-mosses form the principal constituent of peat, an important source of energy in the past.

On the tops of many of our higher hills, particularly in the west, you can walk over great expanses of a soft, grey carpet of Woolly hair-moss *Racomitrium lanuginosum* with only scattered plants of stiff sedge and mat grass. This monotone changes dramatically in areas where snow lies late into the summer. Here mosses and liverworts are again dominant in the vegetation with the russets and blacks of the liverwort crust over the unstable gravels contrasting with the bright greens and reds of the mosses growing in the melt-water channels. Outcrops of lime-rich rocks also occur high on the hills and are just as important for arctic-alpine bryophytes as they are for lichens and flowering plants.

Curve-leaved bow-moss *Dicrandontium uncinatum* - a fine oceanic moss of rocky woodland and crags

What are mosses and liverworts?

Mosses and liverworts, together with the hornworts, are usually studied together and are collectively called 'bryophytes'. Along with the algae, lichens and fungi, they form part of a group usually known as the 'lower plants' (the fungi are included as honorary members of the plant kingdom). Lichens are sometimes confused with mosses, indeed, one lichen is called 'reindeer moss', but lichens are stable associations of a fungus and an alga and have no relationship with bryophytes. There is little chance of confusing bryophytes with higher plants, except possibly with the unfortunately named club-mosses, common in Scotland, which are really allied to the ferns.

The three groups which comprise the bryophytes are now thought to be only distantly related but share a similar life cycle and simple structure. They are an extremely old group of plants with possible ancestors dating back some 450 million years. They are fairly easily separated from each other with just a little experience.

Mosses have stems with leaves and, with the exception of the bog-mosses (*Sphagnum*), have a superficially uniform structure which masks a considerable complexity and variation. Some mosses like the Common haircap *Polytrichum commune* can have shoots up to 80 centimetres tall and the aquatic Greater water-moss *Fontinalis antipyretica* can be over a metre long. These giants contrast with tiny ephemeral species like the earth-mosses (*Ephemerum*) and bladder-mosses (*Physcomitrium*) which are only a few millimetres high. Both *Polytrichum* and *Physcomitrium* form cushions or turfs of upright stems but many mosses creep along the ground, clambering over rocks, trees and other plants. These carpets are a feature of our woods and heaths.

There are two very different forms of liverwort. Like mosses, leafy liverworts have stems and leaves but the leaves are arranged differently on the stem, often with two leaves placed laterally and a row of smaller 'underleaves' below. There is a much greater variety of leaf shape and ornamentation than in mosses and a very different fruiting body.

3

Again there is a huge variation in size and shape amongst the leafy liverworts from large and relatively simple plants like Taylor's flapwort *Mylia taylori* to the tiny, but complex, Toothed pouncewort *Drepanolejeunea hamatifolia*. The other group of liverworts lacks the differentiation into stems and leaves and consists of a strap of green tissue. These species are usually called 'thalloid' liverworts. The large size and net-like surface pattern of the Great scented liverwort *Conocephalum conicum* make it one of our more familiar thalloid species, in contrast to the small rosettes of Crystalwort (*Riccia* species), frequent but mostly overlooked on bare soil.

Hornworts look very similar to thalloid liverworts but differ in a number of technical ways and the fruiting body, which is green and long-lived, is completely different. The two groups are only distantly related. Hornworts are easily overlooked, usually growing on disturbed soil in 'weedy' places in gardens, ditches and arable fields.

Bryophytes reproduce sexually, a process which requires at least a film of water, and produce a 'sporophyte', which consists of a stem with a capsule at the top containing the spores. The form of this fruiting body differs in mosses, liverworts and hornworts and provides the principal means of distinguishing them.

Moss sporophytes are very varied and provide a useful means of identifying species. The pictures in this book give some idea of the range of size and shape. The moss capsule is a complex and beautiful structure when seen through a microscope, having evolved various means of controlling spore release. Liverwort sporophytes are short lived and look very different to those in mosses as the picture of Overleaf pellia *Pellia epiphylla* shows. Hornwort fruiting bodies are different again, being green and horn-shaped, hence the name. They are long-lived and have no capsule but split lengthways to release the spores.

In a large number of mosses and liverworts sexual reproduction seems to be a rare event, and many species have evolved forms of 'vegetative reproduction' giving clones of the parent material. Most green tissue of mosses and liverworts can produce new plants and the simple fragmentation of stems or leaves is thought to be an important mechanism for dispersal. For this reason, trying to 'wire-brush' mosses off an old wall is doomed to failure! Many species take this dispersal tactic a good deal further. The special mechanisms that have evolved include deciduous or fragile leaves, the production of small particles of plant tissue called 'gemmae', either in special organs or attached to the leaf surface, and even the growth of tiny plantlets called 'bulbils' in the angle where the leaves join the stem.

Taylor's flapwort *Mylia taylori* - a large and common liverwort in western woods

Overleaf pellia *Pellia epiphylla* - a very common thalloid liverwort, seen here with abundant sporophytes

Bird's-foot wing-moss *Pterogonium gracile* - a handsome moss of basic rocks in the west

Dwarf bladder-moss *Physcomitrium sphaericum* - on bare soil in the drawdown zone of the reservoir at Harperigg

Uses of mosses and liverworts

In the modern world the economic uses of mosses and liverworts are not immediately obvious. But they do have an important and subtle ecological role, both in water-retention and stabilising mobile surfaces like landslips, scree slopes and sand dunes. Examination of almost any handful of moss will also reveal another important ecological role - that of providing shelter and humidity for a remarkable diversity of invertebrates. This is an integral part of the food web in many of our important habitats.

In the past, mosses were used as an easily obtainable and multi-purpose packaging material. Wads of mosses were used to line pits, probably for the storage of vegetables, and also as stuffing for bedding, along with bracken and straw. 'Ötzi', the 5,000-year-old body of a hunter found in a glacier in the Tyrol, had large quantities of moss stuffed inside his clothing. This was possibly for insulation but more probably as a packing material for his food and this kind of use was probably widespread. It is likely that moss has a long tradition of use for 'personal hygiene', a tradition kept up by desperate hill-walkers today. Moss was also used as caulking (a filler) by Bronze and Iron Age boat-builders and this use continued up until the early 19th century in northern Scotland. Similar use was made of moss to pack the walls of stone houses, particularly near the chimney to keep out the wind and prevent the heat from setting fire to the wooden frame. In the Victorian era, moss was used around the base of potted plants and in flower arranging when 'moss-gathering' was a small industry. This industry is now showing signs of a significant revival with 'moss' being harvested from plantation woodland for the horticultural trade.

The most important mosses in economic terms are *Sphagna*, the bog-mosses. In Scotland the dead remains of *Sphagna* are the main constituent of the peat which covers some 10% of our land. Peat was an important source of fuel over much of the Highlands and peats are still cut as fuel in much of the north and west. Peat has formed over thousands of years when the rate of plant growth has exceeded that of decay. The rate of decay was slowed by the water-logged nature of the ground and the acidity of the ground-water. Peat accumulation started as the climate became wetter soon after the ice retreated some 10,000 years ago. This sometimes reaches 80 centimetres per 1,000 years.

When you walk over a bog, you are walking on water. Over 85% of the bog, by weight, is water, much of it held in the special dead and empty cells which make up most of the leaf of the *Sphagnum* plant. This characteristic is shared by peat which is often 90% water before it is dried out for burning. Then it has an energy output of about half that of coal but much more than wood. The water-retaining quality of *Sphagnum* peat is also much prized by the horticultural industry. Peat is still extracted by milling off the surface layer and is sold in garden centres across the country.

This absorbent quality, plus its mild antiseptic properties, also gave rise to a little-known industry in Scotland, the harvesting of living *Sphagnum* for use as wound dressings. The value of *Sphagnum* for covering wounds has been known for centuries. But it was not until the early 20th century that there was widespread commercial production, reaching its peak in the First World War, when some one million *Sphagnum* dressings per month were used by the British forces. *Sphagnum* was harvested, cleaned and dried and sent off in bales to factories where it was sewn up in fine muslin and sterilised. The centre of production in Scotland was in the Borders and Dumfries and Galloway, but collecting went on over much of the country.

Names

Very few mosses and liverworts have names that are in common usage and the different species are usually referred to by their Latin names by botanists and just as 'moss' by almost everyone else. The name 'liverwort' comes from the supposed resemblance of thalloid liverwort to the liver and has its origin in the 'Doctrine of Signatures' of the old apothecaries, whereby similarity of shape was believed to confer healing powers. A list of common names for bryophytes has now been compiled, all but a few being recent inventions.

The Gaelic term *mòine,* a mossy place or a mire, often appears as a place name on maps but there is also a generic term for moss, *còinneach,* and it seems likely that this usually referred to *Sphagnum*, which was important because it was a useful plant. The red forms of *Sphagnum* were called *còinneach dhearg,* red moss, and had some therapeutic value in the Western Isles:

"When they are in any way fatigued by travel or otherwise, they fail not to bathe their feet in warm water wherein red moss has been boiled, and rub them with it on going to bed."
Martin Martin, A description of the Western Islands of Scotland, 1703.

Why Scotland has so many mosses and liverworts

It will come as no surprise to learn that our damp climate is a principal factor in producing our diverse moss and liverwort flora. It is not the absolute amount of rainfall that is crucial, but the absence of long periods of drought and this, coupled with relatively frost-free winters, make our western seaboard a paradise for bryophytes. Added to this oceanic climate is a diverse geology, giving a mixture of rock textures and chemistry, a recently glaciated landscape, with much exposed rock in crags and screes, and turbulent burns in deep ravines to maintain the humidity. This type of climate and landscape is quintessentially Scottish and is shared by very few places elsewhere in the world. The rocky terrain has also meant that tracts of woodland have been left relatively undisturbed, particularly in ravines, and these old woodlands are the heart of our oceanic flora. Our position on the western fringe of Europe also means that much of the best habitat for bryophytes has been only marginally affected by pollution.

The wet, cool climate has also given us vast areas of mire and its associated bryophyte species. Our mountains are high enough to have an alpine flora with some of the better hills being as rich as many places in the Alps, particularly where patches of snow persist well into the summer. We are also far enough north to have some species whose affinities are with the tundra. In the drier east, particularly in the straths of the Spey and the Dee, there are areas with more 'continental' species, especially where there are expanses of dry, block scree within the Caledonian pine forest. Some southern species which prefer dry, sun-warmed slopes, can thrive in those places in the east where there are south-facing coasts, as on the Black Isle, in Fife and in Berwickshire. Again in the east, where there is still much arable farming, there is a whole range of 'weedy bryophytes' - short-lived species which are most easily seen in stubble fields before they are ploughed again. All these different niches exist in close proximity, giving our small country its extraordinary moss and liverwort diversity.

Falls of Acharn, Perthshire - a ravine cutting through woodland, where scarce liverworts may grow

The Atlantic woodlands

The steep and rocky west coast of Scotland, riven with deep ravines and regularly buffeted by moisture-laden winds straight off the Atlantic, is an ideal habitat for bryophytes. The overwhelming impression in these magical places is of a thick carpet of green festooning rocks and trees. This carpet is usually a mixture of mosses and liverworts, but it is the 'oceanic' liverworts that are of particular interest here, as a number of them have a very limited distribution in Europe and some are globally rare. Two of the most characteristic species, typical of the best oceanic woodlands, are Spotty featherwort *Plagiochila punctata* and Prickly featherwort *Plagiochila spinulosa*, which often grow together with Wilson's filmy fern *Hymenophyllum wilsonii* on rocks and trees. Both of these species, although locally abundant in the west of Scotland, have a very restricted distribution elsewhere, being limited to the extreme western seaboard of Europe and the Atlantic islands such as the Azores and Madeira.

Also occurring with these two species in the best woodland is Deceptive featherwort *Adelanthus decipiens,* which reaches its northern world limit in Scotland. Outside the British Isles it occurs only in north-west France and Spain.

Where there are ravines cutting through the woodland, the bryophyte flora becomes even richer, although you have to 'think small' to find the best plants. The speciality here is a community of tiny liverworts growing on the steep faces of large rocks, and occasionally on trees, close to the river or burn, where humidity remains constantly high and frost is rare. Some of these species such as Toothed pouncewort *Drepanolejeunea hamatifolia* have a very limited distribution worldwide, while others such as Hutchin's hollywort *Jubula hutchinsiae* var. *hutchinsiae* and Brown scalewort *Radula aquilegia* are European endemics with their world headquarters in the west of Britain.

Western featherwort *Plagiochila atlantica* - a large liverwort locally abundant on rock and trees in the west

Toothed pouncewort *Drepanolejeunea hamatifolia* - a tiny liverwort on rocks and trees where the humidity is always high

Brown scalewort *Radula aquilegia* - an oceanic liverwort of damp rocks

Deceptive featherwort *Adelanthus decipiens* - reaching its northern world limit in Scotland

The Atlantic heaths

In western Scotland there is a series of oceanic heaths which have a very interesting moss and liverwort flora. These bryophyte-rich heaths are usually limited to steep and rocky slopes on the north or north-east side of the hills, with heather dominant on the lower slopes and blaeberry becoming more frequent higher up. Under the dwarf shrubs is a thick mat of bryophytes in which species of *Sphagnum* are common. But it is the large, leafy liverworts that provide the interest, often forming swelling cushions of brown, orange and red.

Disjunct distributions of liverworts of the Atlantic heaths

Species	Distribution outside Scotland
Lindenberg's featherwort *Adelanthus lindenbergianus*	W. Ireland, E. Africa, Madagascar, Central and S. America, Antarctica
Alpine notchwort *Anastrophyllum alpinum*	E. Himalaya, W. China, Aleutian Islands
Donn's notchwort *Anastrophyllum donnianum*	Faroes, S.W. Norway, Tatra, W. Tibet, Himalaya, W. China, Alaska, W. Canada
Joergensen's notchwort *Anastrophyllum joergensenii*	S.W. Norway, W. China
Arch-leaved whipwort *Bazzania pearsonii*	W. Ireland, E. and S.E. Asia, N.W. America
Northern prongwort *Herbertus borealis*	S.W. Norway
Wood's whipwort *Mastigophora woodsii*	W. Ireland, Faroes, N.W. America, Himalaya, W. China, Taiwan
Carrington's featherwort *Plagiochila carringtonii*	W. Ireland, Faroes, E. Himalaya, W. China
Cloud earwort *Scapania nimbosa*	W. Ireland, Himalaya, W. China
Bird's-foot earwort *Scapania ornithopodioides*	England, Wales, W. Ireland, Norway, Faroes, Himalaya, W. China, Japan, Taiwan, Philippines, Hawaii

There are a number of uncommon species here which have an extraordinary world distribution in a number of disjunct (widely separated) localities.

The best stands of this heath are in the north-west with good examples on the remote Sutherland hills and the mountains of Torridon, where Beinn Eighe has the only British population of Northern prongwort *Herbertus borealis*. What all these sites have in common is a climate in which long, dry spells are rare and frost is infrequent and there has been little disturbance in recent times. The protection given by the cover of heather or blaeberry seems to be critical and without it these plants are limited to areas of block scree or moderate snow-lie.

Recent survey work in the Hebrides has also revealed that an endemic species of *Sphagnum* also regularly occurs in this community. Skye bog-moss *Sphagnum skyense* was first described from Skye but is also frequent on Rum and in the Harris hills. It often occurs with another near-endemic species Silky swan-neck moss *Campylopus setifolius* which, outside western Britain and Ireland, is only known from one site in north-west Spain.

In European terms, extensive stands of this oceanic heath community are limited to the west of Britain and Ireland, with the greatest extent on the mountains of the north-west of Scotland. We have a particular responsibility for it, though it has been overlooked in conservation terms.

13

Oceanic montane heath on the north slopes of Glas Bheinn, Assynt

Cloud earwort *Scapania nimbosa* - a distinctive liverwort in oceanic montane heaths

The mountains

Scotland also has an extremely rich arctic-alpine bryophyte flora, often occurring in close proximity to the oceanic species, giving plant communities that are unique in Europe. Two special habitats are particularly worthy of mention:

- the hills with outcrops of calcareous rocks, and
- the high hills where snow lies late into the summer.

A high proportion of Scotland's Red Data Book species come from these two habitats.

Scotland has more lime-rich rock on its higher hills than anywhere else in the British Isles and many arctic and alpine species occur only in our mountains. Ben Lawers alone has over 50 nationally rare species, and other important sites include the other Breadalbane hills, Caenlochan Glen, the Ben Alder area, the Ben Nevis massif and, further north, Beinn Dearg and Ben Hope. Ledges and crevices on broken, lime-rich crags are probably the most productive habitat and the 'good ground' is usually heralded by relatively common calcicoles like Frizzled crisp-moss *Tortella tortuosa* and more restricted arctic-alpine mosses such as Red leskea *Orthothecium rufescens*. Winter frosts and the occasional rockfall keep this habitat open so that a variety of species persist here and are not overwhelmed by flowering plants. Special plants include Scottish beard-moss *Bryoerythrophyllum caledonicum*, a Scottish endemic, and Alpine Comb-moss *Ctenidium procerrimum* which, in Britain, is found only on Ben Lawers and in Glen Feshie.

The calcareous scree below the crags is also important and mosses and liverworts are usually abundant, both in the sheltered crevices and on the open surfaces of the rocks. Himalayan fringe-moss *Racomitrium himalayanum*, which has its only European sites in Scotland, prefers the sun-warmed rocks on south-facing slopes. Rarities like Dimorphous tamarisk-moss *Heterocladium dimorphum* and Spinose thyme-moss *Mnium spinosum* prefer the shaded cavities. Another plant of these shaded crevices is Hair silk-moss *Plagiothecium piliferum*, found on Ben Lawers and in Caenlochan Glen but not seen for more than 60 years. The flushes below the crags and scree are also important habitats, being both attractive and having some rare and delightful species. Green spur-moss *Oncophorus virens* prefers the wet grassland at the edge of the flushes but Broad-nerved hump-moss *Meesia uliginosa* is a plant of mossy hummocks within stony springs. Even though these calcareous hills have been the target of bryologists for years, their richness is such that they are still the most likely sites for new finds.

14

Ben Lawers - the famous south-west crags are as important for mosses and liverworts as for flowering plants

Red leskea *Orthothecium rufescens* - a beautiful moss of wet lime-rich rocks in the hills

Alpine comb-moss *Ctenidium procerrimum* - a moss of mountain limestone on Ben Lawers and in Glen Feshie

Himalayan fringe-moss Racomitrium himalayanum - has its only European sites in the Breadalbane mountains

Frizzled crisp-moss *Tortella tortuosa* - a common species of lime-rich rocks

Snow beds - Scotland's Arctic

Walking across the Cairngorm plateau on a fine, windy day in the depths of winter, the snakes of spindrift and the occasional maelstrom reducing visibility to zero, reminds you that snow does not stay put once it has fallen. Most snow that falls is blown onto lee slopes and into gullies where it can form drifts of prodigious depth. These 'snow-beds', on north and north-east facing slopes, can persist right into the summer months and, in the deepest coires and gullies, snow can often last throughout the year. This pattern of snow accumulation and melting means that snow patches usually build up in the same places each year. But it also means that the growing season, for plants underneath them, is short and erratic. Bryophytes and lichens, which can persist and even grow under the snow, have an advantage over flowering plants in these harsh conditions and they dominate in these 'snow-bed communities'.

Blytt's rock-moss *Andreaea blyttii* - a very rare species of flat rocks in areas where the snow lies very late

Where meltwater from the snow patch percolates through the gravel soil, the predominant colour is green from mosses like Ludwig's thread-moss *Pohlia ludwigii*, Northern haircap *Polytrichum sexangulare* and Starke's fork-moss *Kiaeria starkei*. Large patches of the pale, blue-green of Mountain thread-moss *Pohlia wahlenbergii* var. *glacialis* are visible from several hundred metres away. In the best snow-beds, these damp gravels can also have large patches of the liverworts Snow threadwort *Pleurocladula albescens* and Alpine ruffwort *Moerckia blyttii*, both restricted to areas of late snow-lie. On the drier slopes a liverwort crust forms, often much contorted by frost-heave, and the colours here are dark reds and blacks, with the most common plants being species of Rustwort *Marsupella*. This liverwort crust can be very species-rich with a two centimetre square containing ten or more species of moss and liverwort, as well as a number of lichens. On the surface of the rocks, species of *Andreaea*, the 'granite-mosses', are often abundant and include rare species such as Snow rock-moss *Andreaea nivalis* and Icy rock-moss *Andreaea frigida*, on irrigated rocks in burns fed by snow patches, and Blytt's rock-moss *Andreaea blyttii* on flat rocks in areas of very late snow-lie.

There are a number of nationally rare species restricted to this habitat, which is limited in the UK to the higher mountains in Scotland, and even here it probably covers, at most, 1,000 hectares - our own small piece of the Arctic. The largest areas of snow-bed vegetation are in the Cairngorms, which has by far the greatest amount of ground over 1,000 metres, but there are important sites elsewhere, including the Ben Nevis area, the Ben Alder plateau, Creag Meagaidh and, further north, Ben Dearg. The snow-beds have a particular attraction because of their beauty and remoteness, and days spent studying these patches, with snow buntings catching insects on the remaining snow and dotterel nearby, provide some of the most memorable times in the hills.

Snow threadwort *Pleurocladula albescens* - a characteristic liverwort of areas of late snow-lie

Coire an Lochain - home for a large number of snow-bed species in the Northern Corries of Cairngorm

Rivers, springs and flushes

The rocks, gravels and trees in and beside running water are another productive habitat. We have already mentioned the important species that occur on rocks in ravines in our western woodlands but many other interesting species occur in this riparian (riverside) habitat.

Starting up in the hills, the cold water of springs and flushes has its own distinctive flora and the bright greens and reds of the tiny rills are an attractive feature of most coires. One of the most common species is Fountain apple-moss *Philonotis fontana*, often growing with Starry saxifrage, and producing large and distinctive apple-green capsules. There are liverworts here too, one of the most widespread being Cordate flapwort *Jungermannia exertifolia* ssp. *cordifolia* which forms dark-green, swelling cushions in springs. Most of the rarer species, such as Broad-nerved hump-moss *Meesia uliginosa* (see back cover) occur where there is some lime in the run-off so, again, the band of calcareous mountains in the central Highlands, of which Ben Lawers is the best known, includes the most important sites.

Lower down in the burns and small rivers, any stable rock or gravel will have its covering of bryophytes. In the Highlands, the most common species here are Yellow fringe-moss *Racomitrium aciculare*, Rusty feather-moss *Brachythecium plumosum* and Water earwort *Scapania undulata*, and where the water is acidic these species can be overwhelmingly abundant. On slabby rocks at the side of these burns the beautiful Flagellate feather-moss *Hyocomium armoricum* can form large carpets. In both faster flowing burns and more placid water the long fronds of Greater water-moss *Fontinalis antipyretica* are a familiar sight, often one of the first mosses with which people become acquainted.

Even in the lower, meandering stretches of the rivers on the easier gradients in the east of Scotland, where rock surfaces are uncommon, the silt that accumulates on tree trunks in the flood zone provides a home for a few typical species. Probably the most common are Many-fruited leskea *Leskea polcarpa*, River bristle-moss *Orthotrichum rivulare* and Water screw-moss *Syntrichia latifolia* and there are also a few sites for the scarce species Spruce's bristle-moss *Orthotrichum sprucei*. In areas of carr on the Insh Marshes by the River Spey, some of these plants can occur in silt on trees a metre or more above head height, a reminder of winter water levels.

Fountain apple-moss *Philonotis fontana* - a common moss of springs from the hilltops to the valley bottoms

Flagellate feather-moss *Hyocomium armoricum* - an oceanic moss, common on wet, acid rocks

Cordate flapwort *Jungermannia exertifolia* ssp. *cordifolia* - a fraquent liverwort of upland springs and flushes

Greater water-moss *Fontinalis antipyretica* - the waving fronds are a common sight in rivers and lochs

Badanloch Bogs, Caithness - pattern mire in the Flow Country which supports large populations of *Sphagnum* species

Mires

In a wet country like Scotland areas of mire, the boggy places, are a feature wherever you go. In some areas, such as the Flow Country of Caithness and Sutherland, bogs can dominate the landscape and these healthy, active bogs are dependent on the bog-mosses, (*Sphagna*). These remarkable plants not only provide the structure on which the bogs are built but are also beautiful. Each species has its own texture and a distinctive hue, ranging from deep red through orange and ochre to more delicate browns and greens. Though the plants are often lumped together as 'bog-moss', there are 34 different species in Scotland, each with its own particular niche, some preferring the bog pools, others the hummocks and lawns in between, while some are not really bog plants at all.

The large bogs contain some of the most 'natural' vegetation in Scotland and we have a special responsibility within Europe for this habitat. The ground needs to be permanently waterlogged and the particular structure of the *Sphagnum* shoots helps to maintain this degree of wetness. Any kind of drainage can upset this process. Some bog-mosses seem particularly susceptible to any disturbance and the presence on a bog of species such as Golden bog-moss *Sphagnum pulchrum* and Austin's bog-moss *Sphagnum austinii* is a good indicator of a relatively undisturbed mire. There are a number of rare species, such as Olive bog-moss *Sphagnum majus* and Baltic bog-moss *Sphagnum balticum,* occurring on just a few of the bogs. However there is a large area waiting to be explored and it is just possible that these species are not as rare as we think.

Naturally enough, there are interesting species other than *Sphagnum* in these mires. In a few select mires in the east, there are olive-green hummocks of the Red Data Book species Waved fork-moss *Dicranum bergeri* and this area is now the British stronghold for the species. Over in the west of Argyll, a rich valley mire is the habitat for one of Europe's rarer liverworts, Marsh flapwort *Jamesoniella undulifolia*, which is also included on a provisional list of the world's threatened bryophytes. This liverwort creeps through *Sphagnum* hummocks with just the apex of the shoot visible and it has been lost from a number of sites in Britain as mires have been drained. Bogs are also home to the dung mosses (*Splachnum* species), growing on the dung of sheep, cattle and deer. The ripe capsule has a smell that attracts dung-flies and, unique amongst mosses, the spores are sticky and adhere to the flies and hitch a lift to the next pile of dung.

Golden bog-moss *Sphagnum pulchrum* - an uncommon bog-moss of wet areas in undisturbed mire

Warnstorf's bog-moss *Sphagnum warnstorfii* and Rigid bog-moss *Sphagnum teres* - two species of bog-moss which prefer more base-rich mire

Round-fruited collar moss *Splachnum sphaericum* - a moss common on sheep dung in the mire areas

Marsh flapwort *Jamesoniella undulifolia* - a rare liverwort growing through *Sphagnum* in Argyll

Sand dunes

This is not a habitat that most people would associate with mosses and liverworts but the calcareous shell-sand systems that are scattered round our windy coasts have a rich and varied bryophyte flora with a number of rare and threatened species. The most interesting species occur where the sand is irrigated by fresh water, either in dune slacks or where small burns run through the dunes. However, the sand also needs to be mobile enough to prevent the establishment of a complete sward of flowering plants. There are a number of very rare *Bryum* species which are limited to this habitat in Britain. *Bryum* species are not the easiest of mosses to identify but two species, Blunt bryum *Bryum calophyllum* and Baltic bryum *Bryum marratii* are more straightforward as they have characteristically blunt leaves and distinctive capsules. Both have much of their UK population in Scotland. Petalwort *Petalophyllum ralfsii,* a pretty little liverwort which is protected under European legislation, also occurs in this habitat at one site in Wester Ross, where there is a large population.

Petalwort *Petalophyllum ralfsii* (a BAP species) - resembling tiny lettuces in the bare sand in Wester Ross

25

Irish ruffwort *Moerckia hibernica* - a characteristic liverwort of northern sand dune systems

Blunt bryum *Bryum calophyllum* - a rare moss of bare, wet ground on shell-sand

In the north, a number of plants of montane flushes also occur at sea level on the damp shell-sand and these include distinctive mosses such as Short-tooth hump-moss *Amblyodon dealbatus* and Down-looking moss *Catascopium nigritum*, together with the liverworts Irish ruffwort *Moerckia hibernica* and the nationally rare Gilman's notchwort *Leiocolea gilmannii*.

All of these small plants, which exploit gaps in taller vegetation, have big swings in population size from year to year as their dynamic habitat changes. This poses some problems for conservation as it is very difficult to decide if a dramatic decline in a population is part of the natural cycle or something to get really worried about. More generally, this is a habitat which has declined rapidly through the 20th century because of conversion to golf links and afforestation, increasingly intensive agriculture and development for industry and tourism. Most of the best sites are in more remote places such as Invernaver, Sandwood Bay, Islay or the Outer Isles. A few of these species still retain a precarious hold on the links near Edinburgh. It is here on the east coast that most sites have been lost, to golf links in East Lothian and Fife, to scrub encroachment at Tentsmuir, to forestry at Culbin and to industrial development in Caithness.

Sandwood Bay, West Sutherland - the remote dune system is home to a number of uncommon species of moss and liverwort

Drainage and ploughing of a large area of mire for forestry

Muirburn - causes a threat to important bryophyte communities in the west

Threats

As with most other organisms, habitat change is the major threat. Changes associated with global warming may have a severely detrimental effect on our upland bryophyte flora, particularly on the very specialised snow-bed species. Pollution is less of a threat now than it was in the past and we are seeing the re-colonisation of our urban areas by species that are sensitive to dirty air. However, concentrations of pollutants still occur in the snow-pack in the mountains and the effect that this may have on snow-bed species is, as yet, unknown.

Other specific, potentially damaging changes are:
- the construction of hydro-electric schemes,
- roads and drainage for wind farm developments,
- inappropriate afforestation,
- road improvements,
- the loss of wayside and hedgerow trees,
- the infrastructure for skiing and other tourist developments in wild places.

Knowledge of the importance of our flora is now such that sheer ignorance of the conservation value of our bryophytes is no longer an excuse for damage.

The oceanic-montane heath has a very patchy distribution in our western hills, given that the climate and terrain is suitable over much of the west of Scotland. It seems likely that the practice of muirburn and heavy grazing have taken their toll in the past, particularly on the lower ground. Many of the special species in the heath have limited means of spread after damage and one severe fire could wipe out a local population of this globally important community forever. Muirburn is less common in the west now than it once was but it still goes on and is often uncontrolled.

A number of the best areas of oceanic broadleaf woodland in Scotland (and thus in Europe) are threatened by the seemingly inexorable spread of rhododendron. This was originally planted in the gardens of 'big houses' but found the climate in the west to its liking and now totally dominates large areas, often under a canopy of oak trees. It casts a dense shade and its leaf litter is very acidic and decays only slowly so that it eventually smothers the ground flora. In the longer term, this dense covering will also prevent tree regeneration. So, when the current generation of trees dies, there will be no more to replace them.

A number of important woodland areas have schemes to eradicate rhododendron but, at current levels of expenditure and commitment, these schemes are only having a limited impact. We are in danger of losing some of the best moss and liverwort communities in Europe unless we get to grips with this problem.

A large number of mire areas, particularly in southern and eastern Scotland, have been lost to drainage. An even greater loss occured as a result of afforestation when the tax regime encouraged planting of conifers with scant regard for their impact or whether they would ever produce a crop. This has largely stopped now but it has left considerable areas in the Flow Country and Dumfries and Galloway with damaged mires. In some cases it is feasible to remove the trees and block up the drains but the process of recovery will be a long one. The value of our peatlands is now recognised and large tracts are given some statutory protection. There are still local threats from commercial peat extraction, often for the horticultural industry.

The horticultural trade also provides the demand for another potentially destructive industry - the gathering of live mosses and liverworts for floral displays, wreaths and bouquets. Not a great deal is known about the extent of this industry and at least some firms attempt to operate in a sustainable way by gathering only from conifer plantations that are due for clear-felling. However, the indiscriminate nature of ripping up carpets of moss is unsettling, as even within the most uniform of plantation woodland there are interesting habitats and species.

One unusual but very real threat to bryophytes, and to other lower plants, is the declining number of taxonomists. Without this professional core, it will be more difficult for conservation agencies to get the kind of information they need to provide adequate site protection. The strong amateur tradition in Scottish botany still needs the professional scientist to produce the floras and to carry out the research work that underpins the fieldwork.

Damaged peatland - a mire destroyed to provide peat

Rhododendron - the most pressing threat to the mosses and liverworts of our western woodlands

Conservation - how you can help

The lower plants, along with 'difficult' groups of invertebrate animals, have long been the 'Cinderellas' of the conservation industry and this is unlikely to change in the near future. Fortunately, many of the conservation measures aimed at more favoured organisms also benefit mosses and liverworts. So, schemes to increase the amount of broadleaf woodland or to re-establish active mires are generally good news for bryophytes.

What more can be done? Education has an important role to play, in terms of helping land managers to take account of the bryophyte interest on their ground. In more general terms it also encourages the wider public to appreciate both the diversity and the beauty of the mosses and liverworts around them. It is disappointing that the most frequently asked question about 'mosses' is how to get rid of them from lawns and walls!

Specific things that you can do to help are:

Land managers
- control rhododendron,
- allow some grazing in woodland – complete exclusion of grazing may improve regeneration but allows the growth of coarse grasses and bramble which smother bryophytes,

- think carefully about the use of muirburn in areas where oceanic heath might occur.

General public
- stop buying *Sphagnum* peat for your garden,
- enquire of the florist where the mosses in your floral display came from,
- pull up rhododendron seedlings in the wild, hang them up to dry and die.

We are lucky that most of our rarest species have reasonably large and stable populations, albeit on a very small number of sites. Once baseline data on population size and distribution have been gathered, most will only require regular monitoring. A few species do give some concern where the single site could be threatened either by habitat change (Violet crystalwort *Riccia hueberiana* on Bute) or a change in management (Lindenberg's featherwort *Adelanthus lindenbergianus* on Islay). Others are precarious because of the small size of the population.
The rare oceanic liverwort Atlantic pouncewort *Lejeunea mandonii* now seems to be restricted to just three trees in three different ravines in the west.

Schleicher's thread-moss *Bryum schleicheri* var. *latifolium* is now reduced to a single population in one small spring in the hills near Stirling. Here direct action is being taken by SNH and an attempt is being made to re-establish the plant in another flush nearby, from which it disappeared a few years ago.

There is a problem of priorities. The plants which make up our most important communities, ones for which we have international responsibility, are the oceanic species, and most of these are not rare in Scottish terms. Most of them have many sites and often have large populations.

In European terms, most of our conservation effort should be directed towards these important oceanic communities, even though there is little direct threat to individual species. The bulk of our rarest species are montane plants which have their main populations in the Alps or Scandinavia, so the conservation concern here is a national one. This priority needs to be balanced against our international responsibilities for species which may not be rare in Scotland.

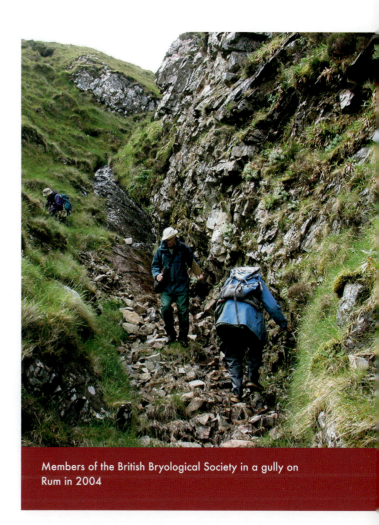

Members of the British Bryological Society in a gully on Rum in 2004

Skye bog-moss *Sphagnum skyense* - a large endemic species of bog-moss

Curled notchwort *Anastroplylum saxicola* - a rare liverwort of dry scree slopes in the Cairngorm area

Dingy grimmia *Grimmia unicolour* - known in just two very different sites in Glen Clova and in Assynt

Schleicher's thread-moss *Bryum schleicheri* var. *latifolium* is known from just one spring near Stirling

Red Data Book and Biodiversity Action Plans

A Red Data Book for British bryophytes was published in 2001 containing 226 species in the various threat categories. Further information is being gathered to assess the status of another 24 species which are known to be very rare. Of the 226 plants, 152 (67%) occur in Scotland and 86 of these have their only British sites here. Many of the most threatened of these species have been the subject of research in the past ten years, funded by SNH, often in conjunction with the Royal Botanic Garden, Edinburgh. This research has resulted in the production of Species Dossiers which attempt to pull together all that we know about each species, especially where it grows, how much of it there is, and how threatened it is.

We are lucky in that many of our rare and interesting species occur in remote and relatively undisturbed habitats where the level of threat is perceived to be low. This is reflected in the sorts of species that were of sufficient concern to need a Biodiversity Action Plan (BAP). There are currently 45 BAP bryophytes, of which 24 occur in Scotland. This relatively low number is symptomatic of the greater pressures on lowland habitats further south.

Green shield-moss *Buxbaumia viridis* (a BAP species) - a rare and ephemeral moss found on dead wood in woodlands in the East Highlands

Finding out more about mosses and liverworts

Most people can recognise 'moss' when they see it, and a few can recognise a liverwort, but beyond this has usually been seen as the realm of the expert. Mosses and liverworts are small and have the reputation of being difficult to identify, but while a few species will always need a microscope to confirm identification, many can be identified at a glance and most can be identified with the aid of a hand lens. Unfortunately, the learning curve is steep at the beginning but, with a little persistence, the most common mosses and liverworts can be readily identified after just a few weeks. The British Bryological Society, a largely amateur society, has an excellent website which can offer assistance and encouragement and this society holds field meetings in Scotland on a regular basis. The website has lots of pictures and can provide links to many more. The Botanical Society of Scotland has a section devoted to lower plants and the Scottish Field Studies Council runs at least two courses on mosses and liverworts each year at its Kindrogan Field Centre, Perthshire.

Members of the British Bryological Society at work

The following books are all useful at different levels.

Crawford, C.L. 2002. *Bryophytes of native woods. A field guide to Common Mosses and Liverworts of Scotland and Ireland's Native Woodlands.* SNH, Native Woodlands Disussion Group, Natural Resource Consultancy, and Eamonn Wall and Co. (Suitable for beginners)

Daniels, R.E. and Eddy, A. 1985. *A Handbook of European Sphagna.* Institute of Terrestrial Ecology.

Hill, M.O. Hodgetts, N.G. and Payne, A.G. 1992. *Sphagnum: A Field Guide.* Joint Nature Conservation Committee.

Paton, J.A. 1999. *The Liverwort Flora of the British Isles,* Harley Books. (The standard liverwort flora)

Perry R.A. 1992. *Mosses and Liverworts of Woodland: a guide to some of the commonest species.* National Museums and Galleries of Wales. (Suitable for beginners)

Smith, A.J.E., 2nd edition 2004. *The Moss Flora of Britain and Ireland.* Cambridge University Press. (The standard moss flora)

Watson, E.V. 3rd edition 1981, reprinted in 1995. *British Mosses and Liverworts: An Introductory Work.* Cambridge University Press. (Suitable for beginners-intermediate)

General

Hodgetts, N.G. and Porley, R.D. (in press) *Mosses and Liverworts.* New Naturalist Series, HarperCollins.

Hill, M.O., Preston, C.D. and Smith, A.J.E. 1991, 1992 and 1993. *Atlas of the Bryophytes of Britain and Ireland, Vol 1, Liverworts and Vols 2 and 3 Mosses.* Harley Books.

Richardson, D.H.S. 1981. *The Biology of Mosses.* Blackwell. (A general introduction to the biology of mosses)

Useful Addresses

British Bryological Society
91 Warbro Road, Babbacombe, Torquay
TQ1 3PS.
www.britishbryologicalsociety.org.uk

Botanical Society of Scotland
Royal Botanic Garden, 20a Inverleith Row, Edinburgh
EH3 5LR.
www.rbge.org.uk

Kindrogan Field Centre
Enochdhu, Blairgowrie, Perthshire PH10 7PG.
Email: Kindrogan@btinternet.com
www.econet.org.uk/kindrogan

Also in the Naturally Scottish series...

If you have enjoyed Mosses & Liverworts why not find out more about Scotland's wildlife in our Naturally Scottish series. Each booklet looks at one or more of Scotland's native species. The clear and informative text is illustrated with exceptional photographs by top wildlife photographers, showing the species in their native habitats and illustrating their relationships with man. They also provide information on conservation and the law.

Amphibians & Reptiles

Although there are only six amphibians and three reptiles native to Scotland, these delightful animals have been part of our culture for a long time. They feature on Pictish stones and in a play - 'The Puddock and the Princess'.
John Buckley
ISBN 1 85397 401 3 pbk 40pp £4.95

Bumblebees

Did you know that Bummiebee, Droner and Foggie-toddler are all Scottish names for the bumblebee? Find out what these names mean and why bumblebees are so special inside this beautifully illustrated booklet. Also discover how you can help the bumblebee by planting appropriate flowers for their continued survival.
Murdo Macdonald
ISBN 1 85397 364 5 pbk 40pp £4.95

Burnet Moths

Unlike many other species of moth, burnet moths fly by day. They can be easily recognised by their beautiful, glossy black wings with crimson spots. Their striking colouring is a very real warning to predators.
Mark Young
ISBN 1 85397 209 6 pbk 24pp £3.00

Corncrakes

Secretive, skulking, rasping, loud, tuneless, scarce . . . all these words have been used to describe the corncrake. But once you could have added plentiful and widespread to the list. Now only a few birds visit Scotland each year. This booklet brings you the latest information on the corncrake and reveals this elusive and noisy bird in its grassy home.
ISBN 1 85397 049 2 pbk 40pp £3.95

Fungi

Fungi belong to one of the most varied, useful and ancient kingdoms in the natural world. Scotland may have almost 2000 larger species with some of the most interesting found in our woodlands and grasslands. This booklet provides an introduction to their life cycles, habitats and conservation. Discover the fascinating forms of earthstars, truffles and waxcaps.
Roy Watling MBE and Stephen Ward
ISBN 1 85397 341 6 pbk 40pp £4.95

Lichens

There are more than 1700 species of lichen occuring throughout the British Isles, and many grow in Scotland where the air is purer. Several different species may be found on a single rock or tree, resulting in lichenologists spending hours in one spot!
Oliver Gilbert
ISBN 1 85397 373 4 pbk 52pp £4.95

Red Kites

This graceful and distinctive bird was absent from Scotland's skies for more than a century. Now with the help of a successful programme of re-introduction, it's russet plumage and forked tail can once again be seen in Scotland.
David Minns and Doug Gilbert
ISBN 1 85397 210 X pbk 24pp £3.95

Red Squirrels

The red squirrel is one Scotland's most endearing mammals. This booklet provides an insight into their ecology and some of the problems facing red squirrels in Scotland today.
Peter Lurz & Mairi Cooper
ISBN 1 85397 298 4 pbk 20pp £3.00

River Runners

Scotland's clean, cascading rivers contain a fascinating array of species. The atlantic salmon is the best known of our riverine species but others, such as lampreys and freshwater pearl mussels, are frequently overlooked but no less captivating. This booklet aims to illuminate aspects of their intriguing and largely unseen lifecycles, habitats and conservation measures.
Iain Sime
ISBN 1 85397 353 X pbk 44pp £4.95

Sea Eagles

This magnificent bird, with its wing span of over 2m is the largest bird of prey in Britain. In 1916 it became extinct, but a reintroduction programme began in 1975. This booklet documents the return of this truly majestic eagle. Production subsidised by Anheuser-Busch.
Greg Mudge, Kevin Duffy, Kate Thompson and John Love
ISBN 1 85397 208 8 pbk 16pp £1.50

SNH Publications Order Form:
Naturally Scottish Series

Title	Price	Quantity
Amphibians & Reptiles	£4.95	
Bumblebees	£4.95	
Burnet Moths	£3.00	
Corncrakes	£3.95	
Fungi	£4.95	
Lichens	£4.95	
Red Kites	£3.95	
Red Squirrels	£3.00	
River Runners	£4.95	
Sea Eagles	£1.50	

Postage and packing: free of charge in the UK, a standard charge of £2.95 will be applied to all orders from the European Union. Elsewhere a standard charge of £5.50 will be applied for postage.

Please complete in BLOCK CAPITALS

Name _____

Address _____

_____

_____

Post Code

Method ☐ Mastercard ☐ Visa ☐ Switch ☐ Solo ☐ Cheque

Name of card holder _____

Card Number ☐☐☐☐ ☐☐☐☐ ☐☐☐☐ ☐☐☐☐

Valid from ☐☐ ☐☐

Expiry Date ☐☐ ☐☐

Issue no. ☐☐

Security Code ☐☐☐ (last 3 digits on reverse of card)

Send order and cheque made payable to Scottish Natural Heritage, to:

Scottish Natural Heritage, Design and Publications, Battleby, Redgorton, Perth PH1 3EW

pubs@snh.gov.uk

www.snh.org.uk

Please add my name to the mailing list for the: SNH Magazine ☐

Publications catalogue ☐